HOW THINGS WORK

THE INSIDE OUT OF CELLPHONES, TV, DRONES, RACE CARS AND MORE! | MACHINERY & TOOLS

 by Tech Tron

First Edition, 2019

Published in the United States by Speedy Publishing LLC, 40 E Main Street, Newark, Delaware 19711 USA.

© 2019 Tech Tron Books, an imprint of Speedy Publishing LLC

All rights reserved.

Without limiting the rights under the copyright reserved above, no part of this publication may be reproduced, stored in or introduced into a retrieval system, or transmitted, in any form, or by any means (electronic, mechanical, photocopying, recording, or otherwise), without the prior written permission of the copyright owner.

All images in this book have been reproduced with the knowledge and prior consent of the artists concerned, and no responsibility is accepted by producer, publisher, or printer for any infringement of copyright or otherwise arising from the contents of this publication.

Tech Tron Books are available at special discounts when purchased in bulk for industrial and sales-promotional use. For details contact our Special Sales Team at Speedy Publishing LLC, 40 E Main Street, Newark, Delaware 19711 USA. Telephone (888) 248-4521 Fax: (210) 519-4043. www.speedybookstore.com

10 9 8 7 6 * 5 4 3 2 1

Print Edition: 9781541968356

Digital Edition: 9781541968554

Hardcover Edition: 9781541968455

See the world in pictures. Build your knowledge in style.
https://www.speedypublishing.com/

CONTENTS

Chapter 1:   How Do Cellphones Work?........................... 6

Chapter 2:   The Anatomy of a Cell Phone.................... 12

Chapter 3:   Cell Phones vs. Smart Phones 22

Chapter 4:   How Does a Television Work? 26

Chapter 5:   The Anatomy of a Television 32

Chapter 6:   How Do Drones Work? 42

Chapter 7:   The Drone and Its Remote Control 48

Chapter 8:   Uses of Drones..60

Chapter 9:   How Do Race Cars Work?...........................66

Chapter 10:   Pocket-Sized Modern Gadgets 86

Chapter 11:   Tech That You Can Make............................106

Experiment 1: Make a battery from
               spare pennies 108

Experiment 2: Build a bike light that works
               using solar power 112

Experiment 3: Build your own periscope 116

We use modern technology everyday. The impact of modern tech is immeasurable. In fact, if you ask some people today if they could live without their gadgets, the answer would probably be a "no". You see, cellphones, computers, TV and so on have become so embedded in our lives that we have become highly dependent on them. We use them to accomplish tasks. We also use them for entertainment. In this book, we're going to dissect some of the used gadgets today to learn the science of how they work. This way, we'll be able to develop an appreciation for all the hardwork our scientists and engineers did. Let's get started.

CHAPTER 1

HOW DO CELLPHONES WORK?

Cell phones have dramatically altered the way people live and work. Now it's possible to stay in touch with anyone in the world, no matter where you are, as long as you can get a signal on your cell phone. At the current time, it's estimated that over 5 billion people around the world own cell phones. In fact, in developing countries over 90 percent of the phones used are cell phones, since these countries don't have the infrastructure of large networks of landlines. Landlines are the types of lines you need to use telephones that are wired to the wall. Cell phones are also called cellular phones or mobile phones. Today, most of the cell phones used around the world are also smart phones.

An early model wall mounted telephone, in the Champaign County Historical Museum, Champaign, IL

Cell phones are actually telephones that work by radio. Their calls are routed through a network connected to a main network of public telephones. Phones that are attached to landlines work in a very different way than cell phones work.

Imagine that you are sitting at a desk in an office. There is a telephone at the desk and it is attached to a landline. These landlines carry your phone calls using cables that are electrical. The conversation that you're having with your friend who is sitting in another office across town essentially travels a direct route to your friend's landline along wires. That's assuming your friend has a phone that is connected to a landline, instead of a cell phone.

Of course, this is a simplification since there are satellites as well as fiber-optic cables and other ways that these landlines connect throughout the network, but essentially your voice goes through your phone from one landline to another to have a back-and-forth conversation with your friend on his or her landline phone.

Landlines carry your phone calls using cables that are electrical.

Cellphones use radio waves to transmit and receive sounds

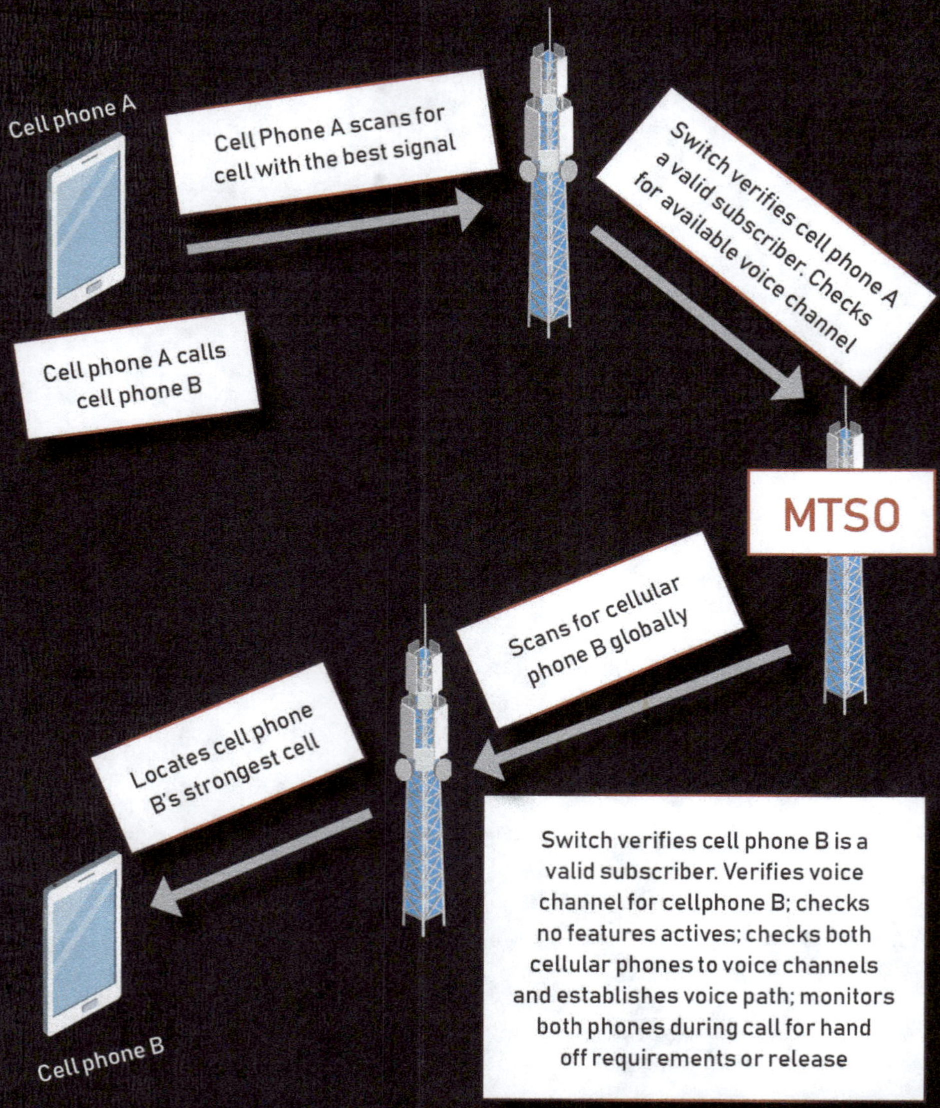

Cell phone A

Cell Phone A scans for cell with the best signal

Switch verifies cell phone A a valid subscriber. Checks for available voice channel

Cell phone A calls cell phone B

MTSO

Scans for cellular phone B globally

Locates cell phone B's strongest cell

Cell phone B

Switch verifies cell phone B is a valid subscriber. Verifies voice channel for cellphone B; checks no features actives; checks both cellular phones to voice channels and establishes voice path; monitors both phones during call for hand off requirements or release

A cell phone doesn't need any wires to operate. It uses radio waves, which are electromagnetic, to transmit and receive sounds. Energy from electromagnetic waves is everywhere around us. You can't see these waves, but they're surrounding you when you're sitting at home watching television, walking to school, or traveling on a train. Lots of objects in our surroundings work using electromagnetic waves. Television broadcasts, radio programs, wireless doorbells, and toys controlled by radio all work through processes that use electromagnetic energy. These invisible waves travel across space at light's speed, which is 186,000 miles every second. Cell phones use this energy to send your voice from one location to another.

CHAPTER II

THE ANATOMY OF A CELL PHONE

There's a small microphone within your cell phone. As you talk, your cell phone converts the up-and-down modulating pattern of your voice into electrical pulses. A special microchip within your cell phone takes these pulses and transforms them into numbers. These strings of numbers are then gathered into an electromagnetic signal that is emitted from the antenna or aerial on the cell phone. Your conversation in the form of numbers speeds through the air at light speed until it gets to the closest cell phone mast.

The circuit board of a mobile phone

The mast gets the pulses and then sends them to the base station. This station coordinates all calls within the local region of the network, which is described as a cell. From this station, the calls are sent to their destinations. If a call is made from one cell phone to another within the same network, then it is routed to the station that is closest to the destination cell phone and then eventually to the cell phone you're calling. Calls that started from one cell phone traveling to a cell phone on a different network or to a phone connected to a landline travel a lengthier pathway. Sometimes they must travel a route that connects them to a main network before they can land at their final destination.

WHAT DO CELL PHONE MASTS DO?

It might seem that cell phones aren't any different than walkie-talkies are. With a walkie-talkie, each one contains a sender as well as a receiver. When two people talk, their messages get bounced back and forth almost as if they are two ping pong players sending a ball back and forth over the net. However, there are some issues with radio signals sent in this way. You can only use so many of them before calls from other walkie-talkies start interfering. In order to avoid this problem, cell phones were built to use much more sophisticated technology.

The mast gets captures the electromagnetic signals and then sends them back to the base station.

Walkie-talkies are also called portable radio transceiver sets.

Cell phones can't beam signals very far. This isn't a failure in their design. In fact, they were created this way purposely. The goal is for the cell phone to transmit its signal to the closest mast, which is essentially a very high-powered antenna, and then on to the station. The job of the base station is to discern weak signals from lots of different cell phones and then route these signals so they get to their final destinations. The masts are enormous and are often mounted in places of high elevation, such as hills or tall buildings. If the masts weren't high-powered, our cell phones would need mega-sized antennas and enormous power supplies, which would defeat their purpose since they would be heavy to carry around! Your cell phone's signal is communicated to the closest cell, the one that has the most powerful signal. It achieves this with the smallest amount of power possible. In this way, it retains its battery power as long as it can. This method also reduces the possibility that it will cause interference for nearby cell phones.

WHAT DO THE LOCAL CELLS DO?

Cells can be described as hexagons on a huge hexagonal grid. The phone carriers chop up an urban area into smaller areas that are about 3 miles by 3 miles. So, imagine a regular invisible hexagon inside that square and that's how a cell can be visualized. Each cell contains a main base station that is composed of a tower and a building that holds the radio equipment.

There's a reason these cells need to be used and why cell phones don't communicate directly with each other. Suppose a bunch of people in your location wanted to send and receive calls using the same frequency of radio signals. With all these signals being sent in all different directions, the chances of calls interfering with each other and scrambling each other would be pretty high. One way to possibly solve this would be for different calls to use different radio frequencies so they wouldn't interfere with each other. It would work as if there were a different radio station available for each call.

Main parts of a cell tower

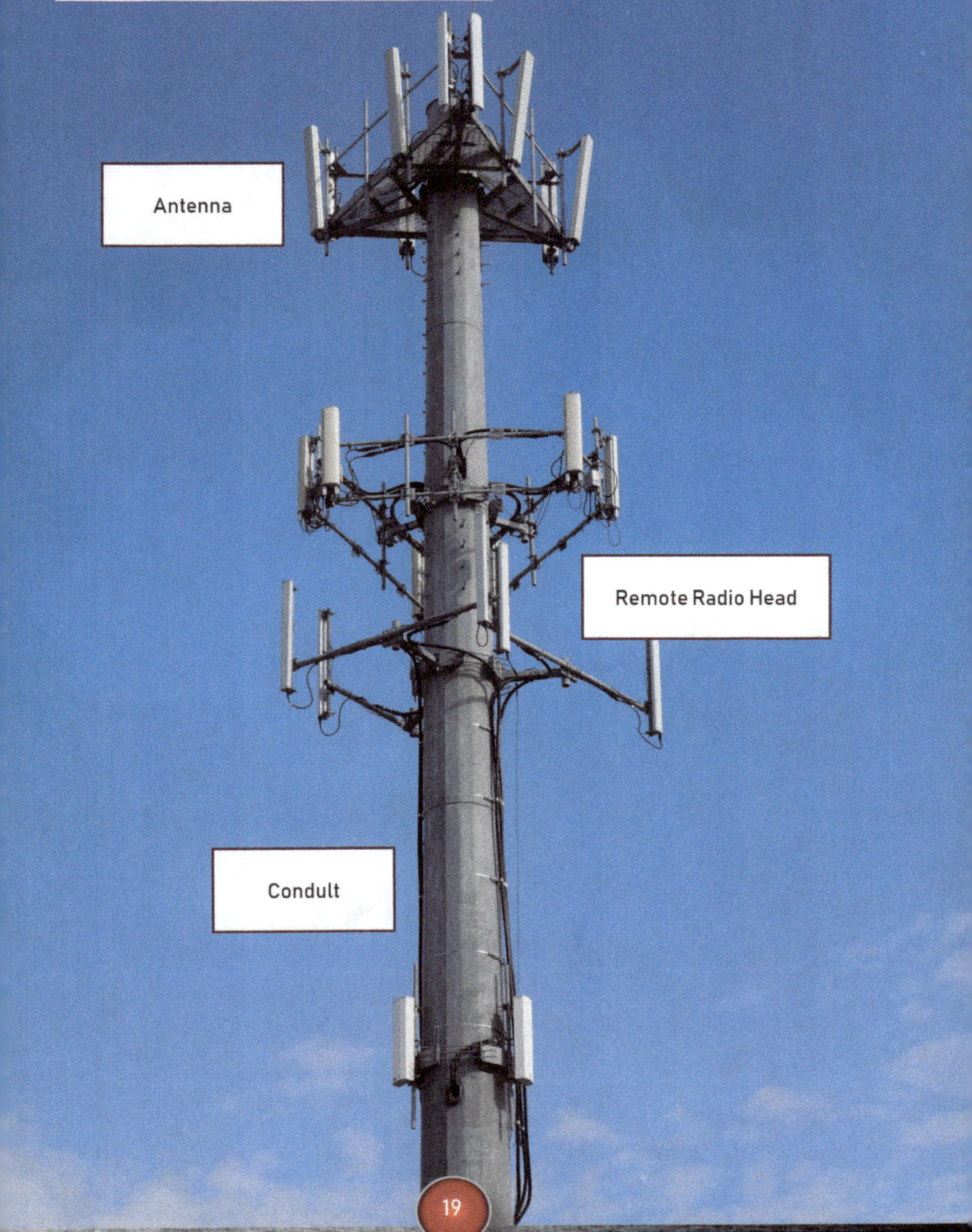

Antenna

Remote Radio Head

Condult

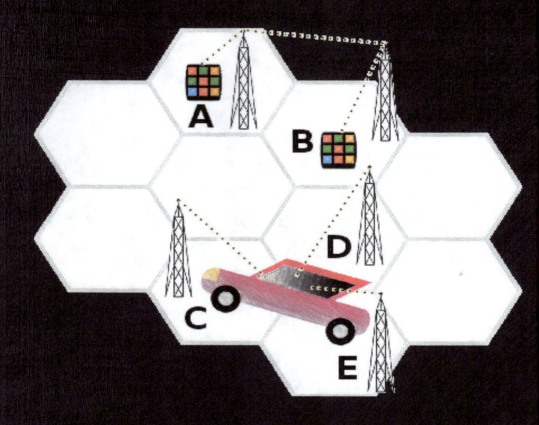

Visualization of a cell tower wherein each cell contains a main base station.

That might work if a small number of people were using their cell phones, but if there were thousands or millions of people using their phones at the same time there would be a need for the same number of radio frequencies, which wouldn't be practical.

A cell site disguised as a tree.

The cells provide an elegant and practical solution to this problem. Depending on the amount of population, phone companies divide areas into cells. If the area is highly populated, the cell size is smaller. In rural areas, where not as many calls need to be handled, the cell size is larger. Because each cell uses the same set of frequencies as the cells that are its neighbors, it enables many more calls to be handled and routed at the same time. The larger the number of cells, the more calls can be routed efficiently.

A cell tower on the roof

CHAPTER III

CELL PHONES VS. SMART PHONES

Most cell phones today are actually smart phones. During the 1990s, cell phones had a lot less functions than they do today. At the beginning, they were mostly used for making phone calls without the inconvenience of wires. Gradually, more and more features were added until the cell phone evolved into a smart phone. Today's smart phones are similar to having a computer in your pocket. They have all the functionality of a telephone, but they also work as digital cameras. They have lots of other functions too. They can play MP3 files like an iPod does, send text messages back and forth, and perform the same tasks as a laptop or tablet using a wide variety of software applications. They also have a global positioning system, which can pinpoint your location and provide you with detailed maps on how to get places. All this functionality has been made possible because today's networks can carry the needed data.

Today, you can also use your smartphone to pay for your purchases.

Unlike telephones that are connected to landlines, cell phones don't need any wires to work. As you talk on your cell phone, the up-and-down pattern of your voice is transformed into electrical signals. A small microchip inside your cell phone changes these electrical signals into a series of numbers. The numbers are broadcast via radio waves, which are electromagnetic energy. Instead of going from cell phone to cell phone, these signals go to local masts and base stations before they are routed to their destination.

An example of a cellphone in the 1990s, which was mostly used for making phone calls but without the inconvenience of wires.

CHAPTER IV

HOW DOES A TELEVISION WORK?

In 1926, John Logie Baird, a Scottish inventor demonstrated a new machine to a group of people in London, England. He had taken everyday objects like a tin for cookies and lights from a bicycle to construct a mechanical television set. At the core of his mechanical television was a quickly spinning disk. It was made from the top of a hat box!

Although the first moment a face was seen on television was historic, Baird's pictures were so unclear that his invention wasn't practical.

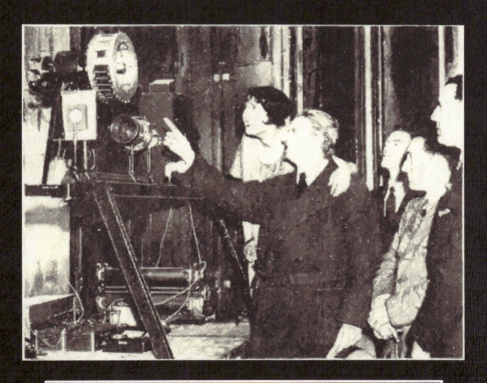

John Logie Baird and his mechanical television.

Several years later, Vladimir Zworykin, a Russian-American inventor made improvements on the cathode-ray tube. The tube was used to show pictures on a screen. By using the cathode-ray tube, he created a television that worked electronically, not mechanically. The earliest televisions had enormous boxes with tiny screens and only showed black and white displays. The television you have at home more than likely uses digital technology instead of the old-fashioned cathode-ray tube.

Television is a complicated invention and many different inventors contributed to its creation. There are three different components that make a television work:

- The television camera: It takes a picture as well as the sound that goes with it and transforms it into a signal that can be sent out on radio waves.
- The television transmitter: It transmits the signals out using radio waves.
- The television receiver: This is your TV set at home. It's able to accept those signals and change them back to the pictures and sound that were recorded by the camera.

Vladimir Zworykin standing in front of the cathode-ray tubes he invented.

Vladimir Zworykin standing in front of the cathode-ray tubes he invented.

Have you ever flipped the pages of a flipbook? Each page has an individual image, but when you flip the pages rapidly, it looks like a moving image. Television does something similar. It transmits still pictures to your eyes so quickly one after the other that they appear to be in motion. There are different rates that are used, but the least amount of pictures it transmits in just one second is 24. That means there are 24 different still pictures being sent to your eyes in just one second when you are watching a television screen! The images are being sent so fast that your brain just blends them into a picture that's moving. Television is similar to a giant flipbook that works electronically.

The evolution of the television from its inception in 1927 until today.

CHAPTER V

THE ANATOMY OF A TELEVISION

HOW DOES A TELEVISION CAMERA WORK?

Light is reflected into our eyes so we can see things. If you were taking a photo with an old-fashioned camera, you could snap a picture by using film that is sensitive to light. It would capture the light on film that would show how an object or person or scene looked exactly at the moment you took the shot. A digital camera does this too, except it doesn't use film. It just captures and saves that moment electronically. A television camera captures new pictures at a rate of at least 24 pictures every second to give the illusion of a picture that's moving.

The television camera takes a picture as well as the sound that goes with it and transforms it into a signal that can be sent out on radio waves.

On a digital camera, you look at a display to see how your finished shot will look. A television cameraman doesn't look through the lens of the TV camera. Instead, he looks at a screen that shows an image of what the lens is "seeing."

HOW DOES A TELEVISION CAMERA TAKE A PICTURE?

Suppose that you wanted to copy a masterpiece that you saw on an art gallery wall. You could take a sheet of paper and divide it up using squares. This grid would help you to concentrate on each section of the piece of art that you wanted to copy. You could work from the top to the bottom or from the left side to the right side.

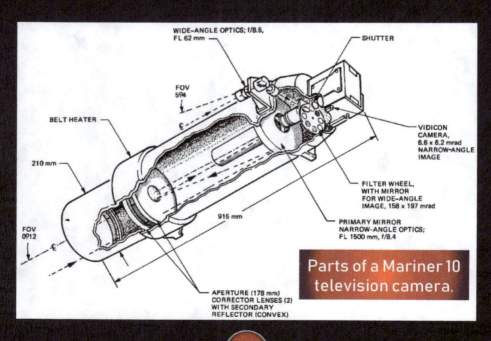

Parts of a Mariner 10 television camera.

In order to transform a picture into a radio signal that can be broadcast, a television camera copies the scene a line at a time. There are detectors inside the camera that sense the light across the picture and scan it line by line just like your eyes scan the text in a book. The process transforms each image into 525 lines of light that have varying colors. The images are beamed using radio waves that travel through the air and arrive at your home television as a video signal. Along with these images, microphones record the sounds that go with the images. The audio signal is sent separately with the picture information.

A television camera records a motion picture and then beams them using radio waves until they arrive at your home television as a video signal.

HOW DOES A TELEVISION TRANSMITTER WORK?

Suppose you were playing in the backyard and you wanted your friend in the yard next door to hear you. You would shout at him as loudly as possible so he could hear your voice. Louder sounds make larger waves. Those larger waves can travel a greater distance before they get absorbed by houses, trees, or whatever other objects are in the environment.

Radio waves behave in a similar way. To make strong radio waves that have the ability to carry pictures as well as sounds, you need a very strong transmitter. This transmitter would have to send the signal from the TV station to someone's home television set. The transmitter is like an enormous antenna that is located on the peak of a hill so it can send the signals out over long distances.

Aerial view of Vilnius TV Tower

However, your home might not receive television signals in this way. There are two other ways you could be receiving your signals. If you have cable in your home, your television pictures are transmitted via a special cable called a fiber-optic cable. These cables are placed under the street. If you don't get your television signals that way, they might come to you by satellite. With satellite television, the picture that you see has been sent out to space to a satellite that's orbiting the Earth. Then it's transmitted back to your satellite dish so your television set can receive it.

Laying of fiber optic cable in Hanover, Germany.

Originally, most television broadcasts were sent with analog signals. The signals were sent in a wave that was moving up and down. However, lots of countries are in the process of switching to television that's broadcast digitally, similar to the way that digital radio is broadcast. The signals are sent in a numerical code. More programs can be transmitted this way and the quality of the picture is actually better because the signals aren't at risk from interference.

Analog **Digital**

Visual comparison of analog and digital signals in television

HOW DOES A TELEVISION RECEIVER WORK?

It really doesn't make any difference which way your television set receives the television signal. It can come from an underground cable, an antenna or aerial on the roof, or a satellite dish in your yard. No matter what way it comes in, your television will still use the same process to show the picture to you. It basically does the exact opposite that the television camera does. It turns the signal it receives back to lines that are a faithful reproduction of the original scene that the television crew filmed. Depending on what type of television you have, it might do that conversion in different ways.

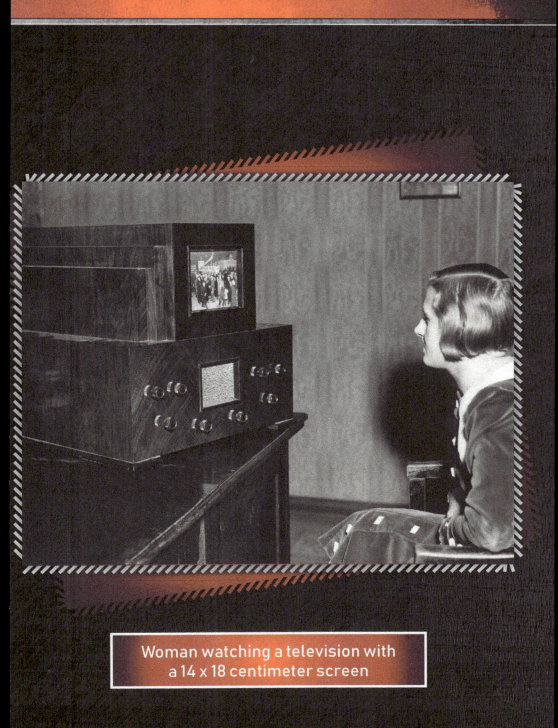

Woman watching a television with a 14 x 18 centimeter screen

CHAPTER VI

HOW DO DRONES WORK?

Drones are unmanned aerial vehicles or UAVs for short. They are operated by remote control. They have many different types of capabilities and have become popular because of their different professional applications. Hobbyists love drones too because they are fun! They seem to fly effortlessly, but there is really a lot of science behind how drones work. The military uses very large drones, but in this book we're going to talk about drones that are hobby size.

Woman landing a drone in her hand

Drones follow the same types of flight principles as other types of remotely controlled vehicles such as model aircraft. However, drones are quad-copters.

HOW DOES A DRONE CREATE LIFT SO IT CAN FLY?

A quad-copter is also called a quadrotor helicopter or quadrotor. It is lifted and propelled in different directions through the use of four rotors as opposed to fixed-wing aircraft like model airplanes. The lift in drones is generated by rotors that are vertically oriented on the craft. In general, they use two sets of identical propellers. One set of propellers turns clockwise and the other set turns with a counter-clockwise movement. So that the user can maintain control over the vehicle, the speed of each rotor can be individually controlled. By changing the speed of the rotors individually, the total desired thrust can be generated. The center of the thrust can be located both side to side and top to bottom to create the needed turning force, called torque.

Different types of quad-copters.

A drone is made from an alloy mainframe that's very lightweight. It has four motors that are run by battery-powered electricity. Each motor has a propeller attached to it. There are speed controllers that are each attached to a motor and are also powered by electricity. There's a mini-computer onboard as well with MEMS, which stands for Micro-Electro-Mechanical-Systems. There are sensors that detect acceleration and there's also a LiPo battery, which is a rechargeable lithium polymer battery.

So, how do all these different components work together to make a drone fly by remote control? Let's look at how the parts work in tandem.

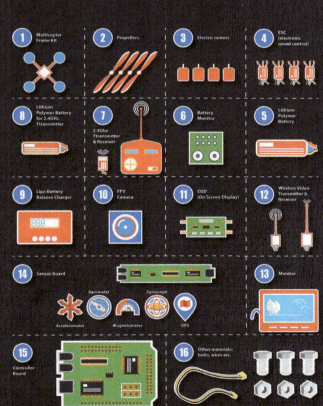

Air drone basic parts

VII CHAPTER

THE DRONE AND ITS REMOTE CONTROL

THE TRANSMITTER

The transmitter is a separate piece of equipment from the drone itself. It can also be an electrical circuit that is found within another electronic device. The job of the transmitter is to allow the user to control the drone remotely from a far distance away. It uses radio signals in the spread spectrum of 2.4 gigahertz. Without the transmitter, you wouldn't be able to control the drone from a distance.

A transmitter and receiver with smartphone preview.

THE RECEIVER

The receiver is another electronic device that is separate from the drone but is needed for the process of operating the drone. It has antennas that are built into it. These antennas pick up the radio signals that have been sent via the transmitter. Those signals are then transformed into pulses of alternating current. The receiver creates information from the current and sends it off to the Flight Control Board of the drone.

THE FLIGHT CONTROL BOARD

The flight control board is part of the drone. After the transmitter starts sending out radio waves and the LiPo battery is plugged into the drone, the transmitter and the receiver start sending communications to each other. These communications are now sent to the Flight Control Board of the drone.

The job of the onboard sensors is to make sure that the drone remains stable even in very windy conditions. This makes it possible for beginners to operate drones more easily than other types of model aircraft.

The onboard sensors ensure that the drone remains stable even in very windy conditions.

The drone board

When the transmitter emits signals to be sent to the drone through the receiver, the computer onboard transmits signals to the controllers that manage the speed of the motors and in turn control the propellers. The controllers can modulate the amount of electrical voltage that the motors receive. This allows the user to manage the speed of each individual propeller. This integrated system is what allows the quad-copter to be easily maneuvered in three-dimensional space.

The flight control board, which is the onboard computer, also has a built-in gyroscope that detects the movement of the drone in three-dimensional space. It's combined with sensors that detect acceleration so that the drone can be maneuvered with precision.

THE BATTERY

The LiPo battery is rechargeable as well as lightweight. It's a pouch onboard the drone that delivers high rates of electrical energy. It has to provide high enough power to make the brushless electric motors turn to move the propellers.

Lithium-ion polymer rechargeable battery (abbreviated as LiPo, LIP, Li-poly) with balancing and main power plugs. LiPo batteries are used in portable electronics, drones and radio controlled models.

THE ELECTRONIC SPEED CONTROLLER

The controller that manages the speed electronically is attached to the battery. It manages the speed of rotation of the electric motor. It does this by adjusting the amperage of the electrical current. This action ensures that the motor is running well and is efficient. The controller has a device built into it that's called a governor. It keeps the rotations per minute or RPM of the motor at a consistent, steady level, no matter what the flight conditions are. It also helps brake the drone if that's necessary.

THE ELECTRIC MOTOR

The drone uses a simple electric motor. The user places electricity in the form of battery power into one end and a metal rod called an axle rotates at the opposite end. This action is what makes it possible to drive the propellers on the drone.

Electronic speed controller

Electric motor

Drone flying over the desert

In other words, on the outside surface of the motor there are magnets. These magnets are latched onto the inner wall, which is called the rotor. The rotor is the spinning part of the motor. There are also magnets that are permanent inside the static part of the motor, called the stator. As electricity travels through the magnets, it creates a field that is electromagnetic. It alternately attracts then repels the magnets inside the stator. This back-and-forth changing polarity of attraction and repulsion is what keeps the motor operating and spinning.

Micro motor, core-less quad-copter motor.

THE PROPELLER

The propellers convert the electric motor's motion into power that lifts the drone skyward. The blades have a special shape that makes the air pressure uneven on the two sides while they are moving. This uneven pressure is what makes the power to lift the drone. Newton's third law of motion and Bernoulli's principle are the underlying physics principles that make it work.

An arm of a small quadcopter drone with a carbon fiber propeller.

THE CONTROL SURFACE

The onboard computer can control each motor's speed individually. For example, if the two propellers on the drone's right side decrease in rotations per minute, the lifting power will be stronger on the left. As a result, the right side will begin to descend quickly and then the drone will drift right. The onboard computer will sense the drift and will increase the rotations per minute automatically if the drifting speed reaches the maximum limit. It will position the drone back into a stable position of hovering, even if the user takes his or her hands off of the controls. This tendency for the onboard computer to correct the motion toward stability is what makes drones easy for beginners to fly.

ON-BOARD DIGITAL CAMERA

Many types of drones have onboard single-lens digital cameras so you can have the drone take photos while it is in flight!

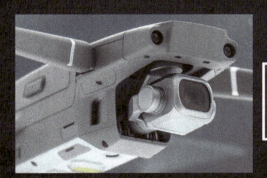

A drone quadcopter with digital camera.

VIII CHAPTER

USES OF DRONES

Large drones have been used by the military for a long time for missions that were dangerous for people and also to collect military intelligence. However, it's only been recently that drones have become small enough and agile enough for small commercial uses and for hobbyists. Inventors have been studying how animals fly to improve the technology for drones and there will be many more innovations in the future.

A 3d rendering of how a drone is used in specialized military operations.

Drones can carry cameras as well as sensors or radio equipment. They can travel places where humans can't easily travel or where it's too dangerous for humans to venture. They can hover like hummingbirds do or swerve around obstacles with the grace of a bat in flight. They can even propel backwards like dragonflies do. They run on clean energy and they're very speedy.

Amazon.com and other companies that depend on fast deliveries are investigating the possibility of using drones in the future to deliver goods and food. Solar drones that could stay airborne for over five years are in the works. These types of drones may be able to provide wireless internet to people who live in remote areas.

Drones with cameras on board are already replacing helicopters to gather news and weather reports. Movie and TV producers are using drones to get expansive landscape views for their productions a lot less expensively than hiring pilot-driven aircraft. Real estate companies are using drones that can take photographs of properties for sale. Farmers can use them to assess the growth of crops. Hobbyists are having a lot of fun flying drones in their backyards. There are even drones you can fly inside your house!

An artist interpretation of the future of drone delivery in online store shipment and delivery service.

Architect flying an inspection drone for an aerial view of a house construction

Of course, with all these drones in the sky people are concerned about privacy. Also, there are safety hazards if people fly drones in areas where they can run into electrical or telephone lines.

Drones are remote-control quad-copters that use four propellers to create the lift needed to fly. They were originally developed for use in the military but today they are small enough and agile enough for commercial and hobby use. Drones have many different components that work together to make them stable flying vehicles. They're a lot of fun to fly!

CHAPTER IX

HOW DO RACE CARS WORK?

Even though they have the same basic parts as ordinary cars, race cars have been developed for only one thing and that is high speed. These cars can get to speeds upwards of 233 miles per hour, even though in a Grand Prix race the speeds might only be around 150 miles per hour.

A race car

CHASSIS

The heart of the race car is its chassis. This is the main portion of the car to which everything is attached. Formula One cars feature a type of construction that's called "monocoque." The word comes from the French language and means that it's a singular shell, which simply means that the body is made out of one piece. Aluminum was formerly used but today some type of carbon fibers that are layered over a mesh made of aluminum or placed in resin are the usual construction materials. These materials produce a car that's very lightweight. It's also very strong and can withstand the tremendous forces acting on the car when it moves rapidly through the air.

A chassis of a sports car

The chassis has a special cockpit designed just for one driver. The cockpit must conform to the Formula One standards, which are rigorous. Each cockpit must have a minimum size and it must also be constructed with a floor that's completely flat. The seat is the only part that is customized for the driver and is created to fit him perfectly so he doesn't move when the car "flies" around the track.

ENGINE

Before the year 2006, F1 cars were run by huge three-liter size engines with 10 cylinders in two banks of 5 each. Then, the FIA, which is the governing body that specifies the rules, changed the specifications to a V8 engine, with a capacity of 2.4 liters. Even though this change meant that the power output of the engines decreased, these engines are still 900 horsepower. As a comparison, the Jetta, a type of Volkswagen, has a 2.5-liter capacity, but only 150 horsepower. You might be able to drive a Jetta for over 100,000 miles before the engine would burn out, but an F1 engine really needs to be rebuilt after about 500 miles of racing. The reason is that the engine has to run at about 19,000 rpm or revolutions per minute. Running the engine at that rate puts a huge amount of tension and stress on its parts.

A powerful engine of a sports car

F1 cars don't use a typical gasoline either. F1 racing car teams use about 50 different types of gasoline blends and additives that boost power are not allowed. Any fuel that is used must meet the approval of the FIA standards.

TRANSMISSION

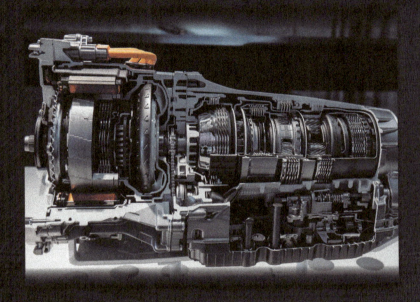

A side view cross section of an automatic transmission gearbox

In order for the engine's power to be transferred to the rear wheels a quality transmission is built into the car. It is bolted to the back of the engine and includes these parts:

- A gearbox—The gearbox can have a maximum of seven gears, but must have at least four forward gears. For a while, gearboxes with six speeds were popular, but now most F1s have seven. They also need to have a reverse gear. The gearbox is attached to a differential.

A red automatic gear stick of a race car

- A differential—The differential gives the car the ability for the wheels to travel at varying speeds when the car is turning. The differential is attached to a driveshaft.
- A driveshaft—The job of the driveshaft is to send power to the cars wheels.

The driveshaft sends power to the car's wheels.

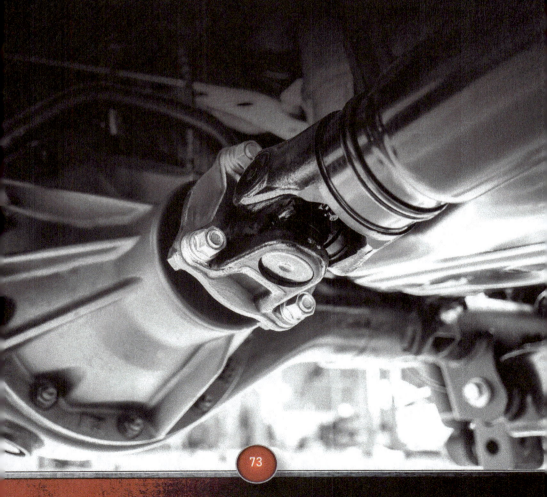

When the driver needs to shift gears in an F1 car, it's very different than shifting gears in an ordinary car with a manual transmission. Instead of the everyday "H-shaped" selector, there are paddles that are situated behind the steering wheel. If the driver wants to downshift, he uses the paddles on one side of the steering wheel, if he wants to upshift he uses the paddles on the other side. Although these cars can be built with automatic transmissions they have been deemed illegal by the FIA. One of the reasons is that the skill that a driver has shifting gears appropriately can be an advantage in winning the race.

AERODYNAMICS

An F1 race car has to have an incredibly powerful engine, but just as important is the overall design of the car. An aerodynamic design is critical to its success at attaining high speeds. To decrease the friction of the atmosphere as much as possible, F1 cars are designed to be low to the ground and quite wide. To increase the car's downforce, it needs four different parts:

- Wings—Wings were first made part of the design in the 1960s. They have a similar function as an airplanes wings except they are supposed to produce the opposite effect. Instead of lifting the car off the track like an airplane, they are designed to push the car down onto the track. This is especially important for when the car is approaching a corner at very high speed.
- Diffuser—Most of the F1 cars have a design that is flat all the way from the nose cone to the line of the rear axle. Most designs have a diffuser. A diffuser is a device that has an upward curve under the engine and under the gearbox. It makes an effect that sucks air up and then sends it out toward the rear of the car.

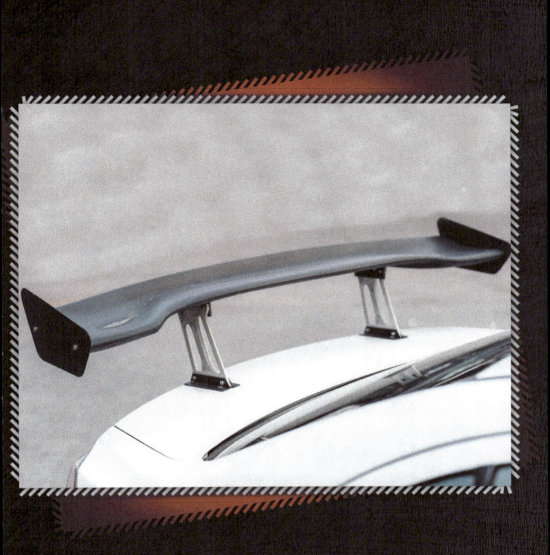

A race car wing

Auto mechanics changing a racing car wheel

- Endplates—Endplates are small flanges on the front wings edges that grab up the currents of air and direct them alongside the body of the car.
- Barge boards—The barge boards grab up the air directed from the endplates and make it go faster in order to create more downforce.

With all these devices to create downforce, the end result is a force of over 5,500 pounds. The car only weighs about one-fourth of the downforce its design creates!

SUSPENSION

A suspension system on an F1 car has the same parts as an ordinary car. Most F1 cars have a type of suspension system called a double wishbone. Before the start of any race, the team responsible for the driver of a particular car will adjust and tweak the settings to make sure that the car is able to stop as well as take corners safely. It could mean life or death to the driver.

BRAKES

The brakes on an F1 race car are similar to those in ordinary cars except that they must be able to halt the car when it is traveling at very high speeds of 200 miles per hour or more. When the brakes are applied, they get very hot to the point where they actually glow red. Special discs made of carbon fibers and specially designed pads are used to reduce the wear on the brakes. They are effective up to temperatures over 1300 degrees Fahrenheit and are still very lightweight.

Holes positioned around the disc's edge make it possible for the heat to diffuse quickly. In addition, the cars have intakes for air that are on the outer part of each wheel's hub that are designed to cool the brakes down rapidly.

The brake system of a sport car

A car tire of a sports car

TIRES

In some ways, the tires are the most important part of an F1 car. They are the only part of the car that has contact with the road. Every part of the car can be working at maximum efficiency and effectiveness, but if the F1's tires don't travel well the driver and car can't win the race. Just like all the other parts of an F1 car, the approved standards are very specific. There are minimum and maximum widths for the tires as well as specifications for the tire grooves. The tires are composed of a type of rubber that is very soft so it will hold onto the road as much as possible. To perform at their best, the tires must be heated up before the race starts. These tires only last for 120 miles or so.

STEERING WHEEL

The wheel used for steering in an F1 car doesn't look like the one in an ordinary car at all. It looks more like a command hub with tons of buttons and switches. The driver can manage all aspects of the car with just a touch. The steering wheel is half of the width of the wheel in an everyday car.

The FIA rules specify that within five seconds a driver must make his escape from the car and only remove the wheel used for steering. To make this possible, the wheel can be detached at its column for a quick getaway.

Formula One race cars are designed for maximum speed and performance. They are built for marathon races and must conform to very specific standards as specified by the governing body called the FIA. They are the fastest cars in the world and can travel up to 233 miles per hour.

A steering wheel

CHAPTER X

POCKET-SIZED MODERN GADGETS

New high-tech gadgets are coming out every year. Some of them are improved versions of products that have been around for quite some time and others are completely new.

ZIP ACTIVITY TRACKER

The Fitbit Company makes many high-tech gadgets to help people track their exercise activity. One of their newest products is called a Zip Activity Tracker. This tiny, high-tech device clips onto your workout clothes and tracks your physical activity in detail. It tracks every step you take, computes the distance, and gives you a summary of the calories you burned when exercising. It then sends the data to your online account. It helps you set your fitness goals by providing you with graphs to show your progress and badges when you reach important milestones.

MINI DIGITAL TOY CAMERA

Lots of different high-tech gadgets take photos, but if you want an inexpensive camera that's almost a disposable, then the Chobi Cam Toy Camera might be just the right gadget for you. It only weighs about 16 grams and is smaller than the palm of your hand. It looks like a yellow wedge of Swiss cheese. It can take high-resolution still photos. It can also record your voice and 45 minutes of video. It has a micro SD (secure digital) card that has 32 gigabytes of memory, so it's pretty powerful for a tiny toy. Next time you give someone the direction "Say Cheese" you can do it with a cheese camera!

A female traveler taking picture with a toy camera

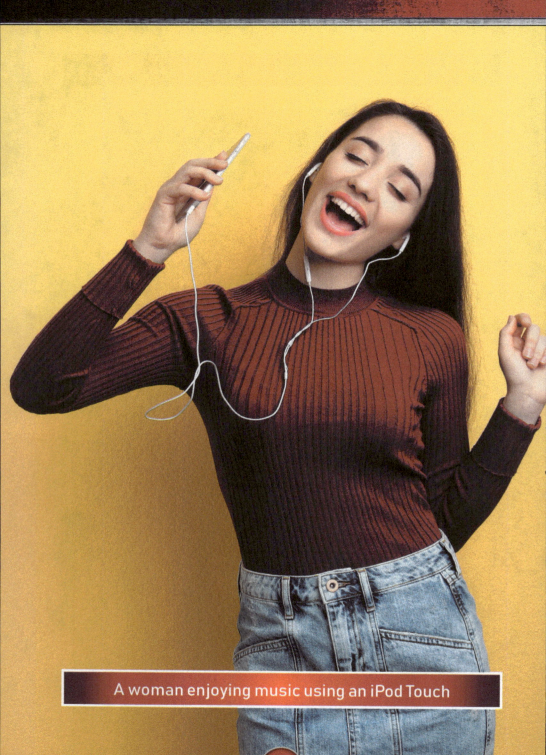

A woman enjoying music using an iPod Touch

IPOD

Apple has created many different kinds of iPods since the product first came out in 2001. The iPod easily fits in your pocket. The most popular version is the iPod Touch because it can connect via Wi-Fi unlike many other models. The iPod started out as a device to listen to a library of a thousand songs that you could download and store to it. Today, it has those features, but it also has a touch screen like an iPhone. In addition to using it to access your music, you can take photos and videos with it. You can create a movie trailer with it by using iMovie.

Because iPod Touch can connect to the cloud via Wi-Fi, you can tap into an unlimited source of storage for photos and videos using iCloud. It can connect to iTunes Match, which is Apple's cloud service for storing your music collection, and also to iTunes Radio, which is an Apple radio service that streams on the internet. You can watch movies and play games or access educational apps on the iPod Touch too.

The iPod Touch also has Siri, which is a personal assistant in the form of a voice. Siri is a type of artificial intelligence. She speaks many different languages. You can ask Siri questions and she will answer. Kids who don't even know how to type words yet can get Siri to help them send messages on the iPod Touch using iMessage.

A boy playing a game on an iPod Touch

A smartstick battery charger

IPOD SPEAKER DOCK

There are lots of different speakers that are built to connect to the iPod. Because the iPod is a small device, when it plays music or podcasts it can use speakers to amplify the sound. There are some speakers that are too big to fit in your pocket, but there is a special speaker dock that looks something like a Lego block. It's very tiny and comes in lots of different colors. It's a fun way to get better sound from your iPod.

SMARTSTICK

There's only one problem with smartphones. They run out of power. When you're at home, you can hook them up to a power outlet and recharge them. However, sometimes they run out of power when you're out and about. The smartstick made by a company called "Pebble" is a pocket-sized battery charger for your smartphone and other handheld devices. It's so powerful that it can give your phone a full charge.

MINI LED BED LAMP

If you like to read or study while you're propped up in bed, then you may want to get a mini bed lamp. This tiny LED lamp has a clamp to attach it to different bed designs. All you need are the batteries for it and it will illuminate your reading area without casting light all over the room. It's also a great lamp for when you are traveling.

MOTION SENSITIVE ALARM

When you're traveling sometimes it's good to have extra protection. There are small, pocket-sized, high-tech alarms on the market that you can attach to a door or window. Not too much bigger than a thumbnail drive, they detect motion and set off a loud alarm if anyone tries to enter your room. They can also be used on public bathroom stalls that don't have proper locks. These alarms ensure that you are safe and have privacy.

Burglar alarm motion sensor installed on the wall

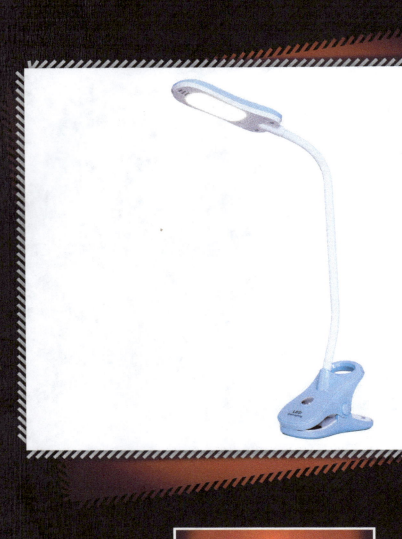

A mini LED lamp

COIN-SIZED TRACKING DEVICE

Are you in the habit of losing things? Then, you might want the new pocket-sized tracking device that's shaped like a coin. All you have to do is attach the tracking device to your valuable item, such as your keys, your phone, or your wallet. You can use the device to ring your phone even if you put your phone into silent mode. You can use the software app that accompanies the tracking device to alert other users to help you find your item too. You can even attach the device to your pet.

SIREN SONG PROTECTION DEVICE

This pocket-sized, egg-shaped, high-tech device attaches to your keychain or other item you carry with you all the time. If you feel that you are in danger, you can sound the alarm, which emits a 130-decibel sound that is deafening, but very effective in getting attention for help. This device is also good for people who have health problems and can't call out for help.

HIGH-TECH WALLET

Did you know that criminals are able to scan your credit cards from a regular wallet? Once your cards are scanned they can use them to buy things or steal your identity. However, there are now sleek, high-tech wallets made out of thin metal or other material that prevents criminals from scanning your credit cards.

An RFID blocking wallet made of metal

A girl cyclist wearing GPS wristband for safety

GPS WRISTBAND

Parents always worry about their children. One device that helps give them peace of mind is a GPS wristband. Some of these high-tech wristbands double as watches also. The GPS (global positioning system) allows parents to know exactly where a child is at any time. These types of devices are also helpful for people who have memory loss. They may get lost and not know where they are and by using the GPS, their caregivers can find them.

HANDHELD GAME DEVICE

Many manufacturers have different handheld devices specifically for playing video games. Some of them contain over 200 games so whenever you want to do something fun, you can just take the device out of your pocket and start playing.

As technology evolves, smart devices get smaller and smaller. New pocket-sized gadgets are being created every year. The power of a desktop computer now fits on a thumbnail drive that can easily sit inside your pocket! There are devices you can use to play your favorite music, to communicate with your friends, to learn new things, or to play games. You can take photos, videos, and create your own movie trailer. Fitness devices help you stick to your health goals by keeping track of the steps you take and the calories you burn. Some devices can help keep you safe in dangerous situations. You can even buy a drone that's small enough to fit in your pocket. Technology has even transformed everyday items such as wallets and keychains.

A girl playing a video game

XI CHAPTER

TECH THAT YOU CAN MAKE

Do you want to be an inventor when you grow up? There are plenty of inventions that you can make on your own with some simple supplies. Here are some ideas to get you started. Make sure you have an adult supervising whenever you do experiments or build inventions.

EXPERIMENT 1

MAKE A BATTERY FROM SPARE PENNIES

In this experiment, you can make your own battery. Your battery will be powerful enough to power a small calculator or light up an LED light.

WHAT YOU'LL NEED

To make this battery, you'll need 14 pennies. The pennies need to be dated no earlier than 1982. That's because pennies before that date don't have enough zinc in them. You'll need some type of acidic liquid, such as lemon juice or vinegar. You'll need some cardboard, a pair of scissors, a screwdriver, and an inexpensive package of zinc washers about the same size as your pennies. You'll also need electrical tape, some aluminum foil, a simple calculator that you don't mind tearing apart, and an LED bulb.

WHAT TO DO

Step 1: Using your screwdriver, take the screws off the back of the calculator so that you can access the battery inside it.

Step 2: Take the battery out and set it aside. You can put it back after you test your new penny-powered battery.

Step 3: Pull the leads out of the calculator's casing. There will be a positive as well as a negative lead.

Step 4: Attach a wire to each lead. You can twist the wire around the end of the lead or use electrical tape to keep them together.

Step 5: Select four of the pennies and four of the washers. Trim out four pieces of cardboard into circular pieces just a little larger than the pennies.

Step 6: Soak the circular cardboard cutouts for a few minutes either in vinegar or lemon juice.

Step 7: On your work surface, place a piece of aluminum foil. Then, stack one zinc washer on the top of it. Next, take one of the vinegar-soaked cardboard pieces and blot the excess liquid off, before placing it on top of the washer. Now, place your penny on the top of the cardboard.

Step 8: You've created a small battery cell. The penny is positive and the zinc washer is negative with an electrolyte, the acid-soaked cardboard in between. You've used a similar process to the one used by Alessandro Volta when he invented the first battery in 1799.

Step 9: You can add cells by constructing another zinc-cardboard-penny stack on top of the one you just did. To power the small calculator, you'll probably need at least 4 cells.

Step 10: Each cell emits about 6/10 of a volt, so if you use 4 cells you'll have about 2.4 volts, which should be enough to power your calculator.

Step 11: Add wires to the top and bottom and use the electrical tape to keep your battery stack together. You can discard the piece of aluminum foil.

Step 12: Now it's time to connect your leads to the appropriate positive and negative leads on your calculator. If you've created your battery correctly, the calculator should power up! Test out a few additions and subtractions to see if your calculator is functioning properly.

Step 13: If your battery ceases working, pull it apart and try soaking the pieces of cardboard in a little more vinegar.

Step 14: Next, follow the same process to create a battery that has 14 cells. This battery should give you 14 x 0.6 volts or about 8.4 volts of power. This should be enough to light up your LED bulb. Place the taller leg of the LED to the top and the shorter to the bottom. Wrap your battery and bulb in electrical tape and see how long the LED stays lit. Some of them have stayed lit for more than a day!

EXPERIMENT 2

BUILD A BIKE LIGHT THAT WORKS USING SOLAR POWER

For this experiment, you'll use an empty deodorant container to make a light for your bike.

WHAT YOU'LL NEED

You'll need an empty deodorant stick and a simple garden light that's powered by solar energy. You'll also need some silicone adhesive as well as an electric rotary tool that can cut plastic. Some hot glue is needed to hold the pieces in position. A screwdriver and a clamp are handy if you want to permanently clamp the light onto your bike.

WHAT TO DO

Step 1: Get rid of any leftover deodorant pieces that are on the stick.

Step 2: Take out the plastic piece of the deodorant stick that pushed the deodorant up to your armpit. There's a piece that holds the deodorant in place and a threaded spindle that pushes it up as the deodorant gets used. If you do this properly, the inside of your deodorant stick should be empty. Keep the deodorant cap. You're basically creating an area for the solar cell to reside.

Step 3: Take your solar light apart. You should see a solar cell. The cell is attached to a circuit board. There should also be a battery pack.

Step 4: There's a small cell near the solar panel piece that is a light sensor. This cell acts like a switch. When it's dark outdoors, the light comes on. When it's sunny out, the light goes off and the battery gets charged from the sunlight. If you're really ambitious, you can wire a switch to this light sensor. Then, you could use the threaded spindle to turn the light sensor on or off. This way you could manually turn on the light when you want to ride your bike in conditions when it's partially dark out.

Step 5: Next, cut a hole in the back of the deodorant case where the ingredients are listed using a rotary tool designed to cut plastic. Make sure an adult is helping with this. You want the solar panel to fit tightly in the hole so it will face out of the body of the plastic deodorant case.

Step 6: The circuit board as well as the battery pack should fit inside the body of the plastic deodorant case. You can use hot glue to make the pieces stay in position until you have it all put together. The LED lights should be positioned where the deodorant used to be.

Step 7: Using some silicon adhesive, seal around the edges where the solar cell fits into the plastic case. The point of doing this is so water won't get in when you're riding your bike in the rain.

Step 8: You can attach your new bike light to your bike's handlebars with rubber bands or a clamp that sits on the handlebar and screws into your new light. Just be careful not to screw too far in because you might damage the solar light's circuit board. Now you can ride your bike to school and the light will charge up for your ride home at night.

EXPERIMENT 3

BUILD YOUR OWN PERISCOPE

WHAT YOU'LL NEED

You'll need two mirrors. The size of the mirrors should be 2 inches by 1 inch but other sizes can also work. You'll need a carton or piece of cardboard that's about 6.5 inches wide and 8 inches in length. You'll also need some scissors, strong glue, and some different colors of paint in green and brown to do a camouflage design when you're finished with the assembly.

Here's a template to help you as you do the assembly.

WHAT TO DO

Step 1: Use the template to draw out your plan for the periscope on the carton or cardboard. The dashed lines mean you should fold along those lines and the lines that aren't dashed are lines that you'll need to cut.

Step 2: Once you have everything drawn out, place the mirrors in the right locations according to the template and then glue them down.

Step 3: Now, fold along the fold lines so that your flat piece becomes a box. If you've done it properly, one mirror should be facing up and the other should be facing down. Use glue to secure the edges so that your carton or cardboard stays box-shaped.

Step 4: Fold each mirror so it's at a 45-degree angle. The mirrors should be parallel to each other. Fold in the side flaps and glue them so the mirrors will stay in position at 45-degree angles and parallel to each other.

Step 5: Now, if you've assembled it correctly, your periscope should work. You can use it to see over a wall or a corner.

Step 6: Use paint to create a camouflage design for the outside of your periscope.

If you think like an inventor, you'll notice there are lots of everyday items that with a little ingenuity can make interesting gadgets. These experiments will give you a starting point. Just make sure an adult supervises, especially when you use electrically powered cutting tools.

A man operating a drone

Visit
www.SpeedyBookStore.com

To view and download free content on your favorite subject and browse our catalog of new and exciting books for readers of all ages.

Printed in the USA
CPSIA information can be obtained
at www.ICGtesting.com
LVHW020252131123
763759LV00004B/24